OCR

A-LEVEL

WORKBOOK

Geography 1

Landscape systems

Changing spaces; making places

Peter Stiff and Andy Palmer

HODDER
EDUCATION
LEARN MORE

Contents

① **This workbook** will help you prepare for the OCR AS and A-level Geography examinations. The layout of subject content closely follows that in the published OCR Geography specification.

② **The workbook** focuses on four topics. The first, second and third are options within the Landscape systems component. You need only study one of them. Changing spaces; making places is a compulsory topic at both AS and A-level.

③ **You should use** this workbook alongside reading the relevant sections of your textbook and checking through your lesson notes. There are also published revision guides that you could consult.

④ **For each topic** you will find a series of review questions that cover the content (enquiry questions and key ideas) set out in the specification. Marks and lines or spaces are given for each question to give you a 'feel' for their relative importance and challenge. These questions build up to, and finish with, exam-style questions. These should be completed within the times shown.

⑤ **In each topic** there are questions built around some form of stimulus material (map, diagram, photograph, statistical table or fact file).

⑥ **Answering the questions** will not only test your knowledge and understanding but will also help you build your exam skills. Answers to all the questions are available at www.hoddereducation.co.uk/ workbookanswers.

⑦ **The AS examination** is 1 hour 45 minutes long, and includes questions on your chosen option in Landscape systems and Changing spaces; making places. You will need to answer a range of questions. Some questions require short answers earning 3 or 4 marks, some are a little longer earning between 6 and 12 marks, while others require extended prose answers for 14 marks, which need to be planned in the same way as essays. Some questions require reference to stimulus materials provided in a resource booklet.

The Physical systems A-level examination (01) is 1 hour 30 minutes long and includes questions on your chosen option in Landscape systems. It also includes questions on Earth's life support systems (see Workbook 2). Some Landscape systems questions require short answers earning 2, 3 or 4 marks, one question earning 8 marks requires reference to stimulus materials provided in a resource booklet and one is an extended prose answer for 16 marks, which needs to be planned in the same way as an essay.

The Human interactions exam (02) is 1 hour 30 minutes long and includes questions on Changing spaces; making places. It also includes questions on Global connections (see Workbook 2). Two Changing spaces; making places questions require reference to stimulus materials provided in a resource booklet, with one earning 3 marks and the other 8 marks. Another question earns 6 marks and there is an extended prose answer that earns 16 marks, which needs to be planned in the same way as an essay.

⑧ **The Geographical debates exam** (03) includes a section in which you will need to make links between an aspect of your chosen options and other areas of the course. Section B (Synoptic questions) could ask you to refer to material from your chosen Landscape system or from the Changing spaces; making places component.

Landscape systems

Option A: Coastal landscapes

How can coastal landscapes be viewed as systems?

Coastal landscapes can be viewed as systems

The coast's highly dynamic environment results from the interaction of atmospheric, marine and terrestrial processes. Coastal landscapes can be viewed as systems, comprising inputs, stores, throughputs and outputs of both energy and material. Systems can exist in a state of equilibrium, when inputs equal outputs. If this equilibrium is disturbed, the system may self-regulate and restore equilibrium; this dynamic equilibrium is an example of negative feedback.

1 **Outline the three types of energy that can be inputs into coastal landscape systems.** *3 marks*

....

....

....

2 **Describe one output of energy and one output of material from a coastal landscape system.** *4 marks*

....

....

....

....

....

3 **What is dynamic equilibrium?** *3 marks*

....

....

....

4 **Distinguish between stores and flows (transfers) in a sediment cell.** *2 marks*

....

....

5 **Explain why sediment cells may not be closed systems in terms of sediment transfer.** *4 marks*

....

....

....

....

Coastal landscapes are influenced by a range of physical factors

Physical factors influence the way geomorphic processes work and how coastal landscapes develop. They can influence the operation of the coastal landscape system by affecting the inputs, stores, throughputs and outputs. The influence of different factors varies spatially and temporally.

Factors include winds, waves, tides, geology and currents.

How much energy waves have, and how they move and break, greatly affects their ability to carry out geomorphic processes.

6 **Explain how winds produce waves.** *3 marks*

....

....

....

7 **Explain why waves break when they reach the shore.** *4 marks*

....

....

....

....

8 **Describe and explain the movement of backwash.** *2 marks*

....

....

9 **Using Figure 1, explain why wave refraction occurs on some coasts and how it influences the distribution of energy along the coast.** *6 marks*

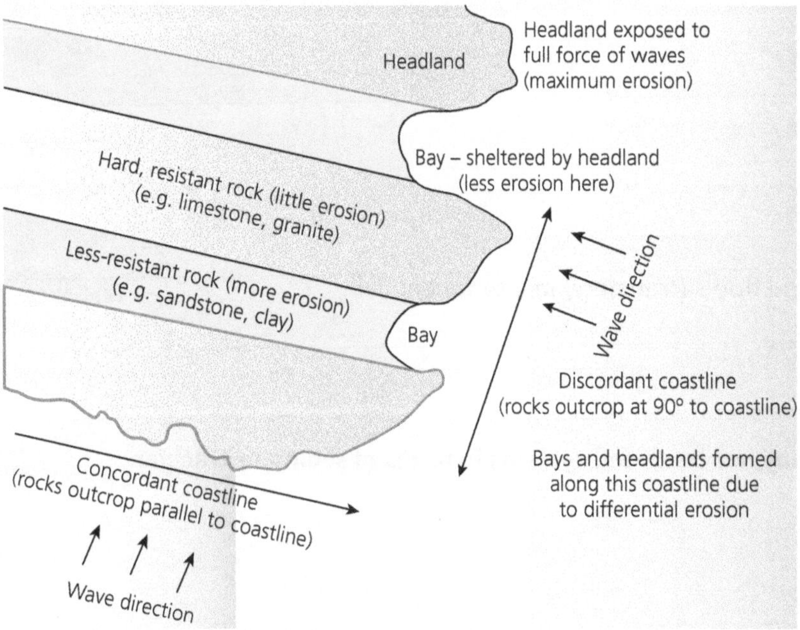

Figure 1

....

....

....

....

....

....

....

10 **Distinguish between spring tides and neap tides.** *2 marks*

....

....

11 **How can ocean currents influence geomorphic processes?** *2 marks*

....

....

12 **Explain how rock resistance is influenced by lithology and structure.** *6 marks*

....

....

....

....

....

....

Coastal sediment is supplied from a variety of sources

Sources include:

- terrestrial: rivers transport eroded sediment to the coast and deposit it, waves erode cliffs, causing undercutting and collapse, longshore drift brings sediment from adjoining beaches
- offshore: constructive waves, tides, currents and winds bring sediment from offshore locations, including sand bars
- human: beach nourishment adds sediment to beaches that have experienced losses

13 **How do rivers gain sediment that can then be added to the coastal sediment budget?** *3 marks*

....

....

....

14 **Explain how waves can produce large amounts of sediment inputs from cliffs.** *4 marks*

....

....

....

....

15   What term is used for sediment transported and deposited by wind?

1 mark

....

16   Describe how longshore drift can help the sediment budget of a beach remain in
equilibrium.

3 marks

....

....

....

17   Why might a sediment budget in a deficit state require human intervention?

2 marks

....

....

18   Briefly describe two ways in which sediment is added to the budget by beach
nourishment.

2 marks

....

....

How are coastal landforms developed?

Coastal landforms develop through a variety of interconnected climatic and geomorphic processes

Geomorphic processes include weathering (physical/mechanical, chemical and biological), mass movement, wave processes (erosion, transportation and deposition), fluvial processes (erosion, transportation and deposition) and aeolian processes (erosion, transportation and deposition).

Climatic processes are interconnected with the geomorphic processes. These include temperature fluctuations, wind and precipitation.

19   What does chemical weathering do to rock?

2 marks

....

....

20   Why might freeze–thaw weathering be of limited importance in temperate coastal
environments?

2 marks

....

....

21   What is the key difference between slumps and linear slides?

2 marks

....

....

22   How much force can breaking waves exert in pounding coastal rocks?

1 mark

....

23 **Describe and explain how sediment can be deposited by waves.** 6 marks

....

....

....

....

....

24 **What is meant by 'flocculation'?** 2 marks

....

....

25 **Explain why aeolian deposition typically occurs inland at coastal locations.** 4 marks

....

....

....

....

Coastal landforms include:

- those that are predominantly the result of erosion, such as cliffs, shore platforms, bays, headlands, geos, blowholes, caves, arches, stacks and stumps
- other geomorphic processes that may also be involved, such as weathering and mass movement (subaerial processes)
- those that are predominantly the result of deposition, such as beaches, spits, onshore bars, tombolos, salt marshes and deltas

However, other geomorphic processes may also be involved, such as longshore drift.

26 **Explain why a widening shore platform may eventually lead to cliff undercutting and collapse ceasing.** 2 marks

....

....

27 **How do geos and blowholes differ in their formation?** 4 marks

....

....

....

....

28 **Explain the influence of subaerial processes on stack formation.** 2 marks

....

....

7

29 With reference to the photograph below, describe and explain the influence of sediment particle size on beach profile.

6 marks

Steep slope angle

Gentle slope angle

....

....

....

....

....

....

....

30 Explain two reasons why the end of a spit may be recurved.

4 marks

....

....

....

....

....

31 Explain how onshore bars and tombolos may be influenced by longshore drift.

4 marks

....

....

....

....

32 Explain the role of vegetation in the development of salt marshes. *6 marks*

Coastal landforms are interrelated and together make up characteristic landscapes

This topic is focused on two case studies: a high energy coastline (such as a rocky coast) and a low energy coastline (such as an estuarine coast) – one of these will be from outside the UK.

For each, you should know and be able to explain:

- the physical factors influencing the formation of its landforms
- the interrelationship of the landforms within the landscape system
- how and why the landscape system changes over time, from millennia to seconds

Note: answers to these questions will depend on your chosen case studies.

33 Explain why your two case study coastlines receive contrasting inputs of energy. *6 marks*

34 Describe the geology of your high energy coastline. *4 marks*

35 Choose two landforms in your high energy coastline and explain how they are interrelated.

4 marks

....

....

....

....

36 For one landform in your low energy coastline, explain the influence of physical factors in its formation.

4 marks

....

....

....

....

37 Describe and explain the long-term (over millennia) changes to your low energy coastline.

6 marks

....

....

....

....

....

....

....

....

How do coastal landforms evolve over time as climate changes?

Sea level changes can be caused by climate changes. Changes in the volume of water in the oceans are referred to as eustatic changes. With increasing global temperatures, sea levels rise because:

- water expands when heated and therefore has a greater volume
- snow and ice stored on the land melt and return to the sea, increasing its volume

The reverse happens with a decrease in global temperature:

- water contracts as it cools and therefore has a smaller volume
- more snow and ice are stored on the land and less is returned to the sea, decreasing its volume

Emergent coastal landforms form as sea level falls

Landforms shaped by wave processes during times of higher sea level are left exposed as sea level falls. These are then left higher, and further inland, in the landscape.

Emergent landforms include raised beaches and abandoned cliffs. These subsequently become modified by subaerial processes. In the long term, sea levels may rise again and these landforms may then become actively affected by wave processes again.

38 **With the aid of Figure 2, explain the formation of a raised beach.** 4 marks

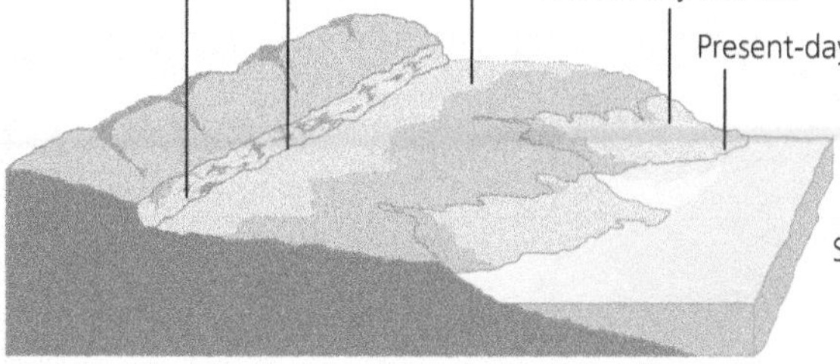

....

....

....

....

....

....

....

....

Abandoned cliff with caves
Raised beach
Present-day cliff line
Present-day beach
Sea

Figure 2

39 **Describe the temperature and relative sea level of one named glacial period.** 3 marks

....

....

....

40 **Suggest how subaerial processes might modify an abandoned cliff during a colder period with lower sea level.** 4 marks

....

....

....

....

Submergent coastal landforms form as sea level rises

Landforms shaped by wave processes during times of lower sea level are submerged as sea levels rise. Submergent landforms include rias and fjords. These subsequently become modified by subaerial processes. Rising sea level may cause a migration of shingle beaches further inland and add larger pieces of sediment to the beach.

41 **Complete the table below to contrast rias and fjords.** 4 marks

	Ria	Fjord
Long profile		
Cross profile		
Water depth		
Formation		

42 **Explain why rias may infill with sediment during sea level rise.** 3 marks

....

....

....

11

43 With the aid of Figure 3, describe and explain the range of projections for sea level change by 2100.   6 marks

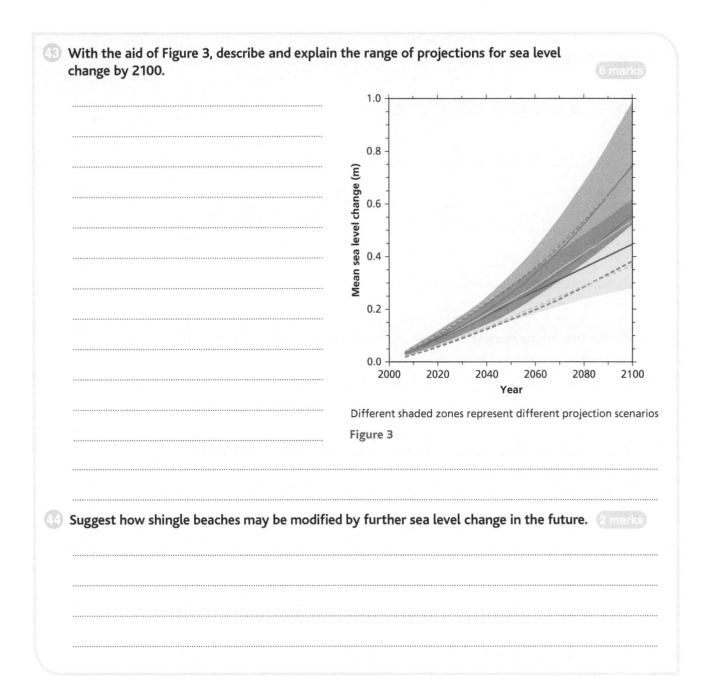

Different shaded zones represent different projection scenarios

Figure 3

44 Suggest how shingle beaches may be modified by further sea level change in the future.   2 marks

How does human activity cause change in coastal landscape systems?

Human activity intentionally causes change within coastal landscape systems

Most intentional changes are the result of intervention attempts to protect the coastline from the effects of natural processes. These often involve hard and/or soft engineering strategies to reduce coastal erosion and/or flood risk.

You will study one coastal landscape that is being managed. You should know and understand:

- the management strategy being implemented and the reason for its implementation, such as use of groynes or offshore dredging
- the intentional impacts on processes and flows of material and/or energy through the landscape system, such as their effect on the sediment budget
- the effect of these impacts in changing coastal landforms, such as beach profiles
- the consequences of these changes on the landscape, such as extension of the coast seawards

Note: answers to these questions will depend on your chosen case study.

45 How and why is your case study landscape being managed? 4 marks

....

....

....

....

46 How has the management strategy affected processes and flows of energy and materials in the system? 3 marks

....

....

....

47 What effects have the changes in processes/flows had on the landscape? 3 marks

....

....

....

Economic development unintentionally causes change within coastal landscape systems

Coasts provide many opportunities for economic activity, such as tourism, trade and fishing. You will study one coastal landscape that is being used by people. You should know and understand:

- the economic development taking place and the reason why it is taking place, such as trade routes, port or tourist resort development
- the unintentional impacts on processes and flows of material and/or energy through the coastal system
- the effect of these impacts in changing coastal landforms, such as beach profiles
- the consequences of these changes on the landscape, such as coastal retreat or protection

Note: answers to these questions will depend on your chosen case study.

48 Describe the economic activity taking place in your case study area. 4 marks

....

....

....

....

49 Explain why this activity is taking place in this location. 4 marks

....

....

....

....

50 Draw an annotated sketch map to show how the changes to the processes/flows have influenced the coastal landscape.

4 marks

Exam-style questions

51 Explain the influence of energy flows on cliff formation.

10   8 marks

....

....

....

....

....

....

....

....

52 Study Table 1, which shows rates of sediment accumulation on a beach in California, USA, in 2017.

Table 1

Months	Rate of sediment accumulation (cm^3/m^2)
Jan–Feb	0.8
March–April	1.1
May–June	3.4
July–Aug	5.6
Sept–Oct	4.5
Nov–Dec	1.4

a Calculate the mean rate of sediment accumulation shown in Table 1. You must show your working.

3 **2 marks**

....

....

b Calculate the median rate of sediment accumulation shown in Table 1. You must show your working.

3 **2 marks**

....

....

53 Study Figure 4, which shows past, present and projected future sea level change.

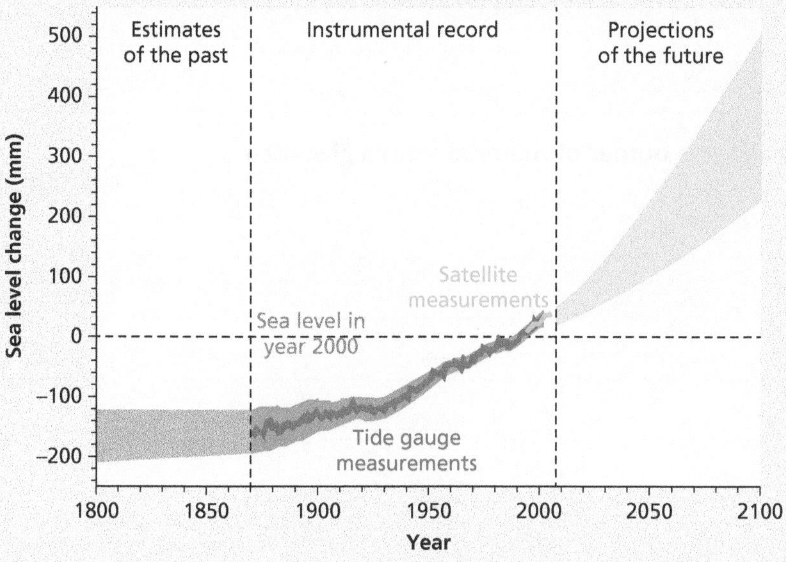

Figure 4

Using data from Figure 4 in your answer, explain how increasing global temperature causes rising sea levels.

7 **5 marks**

....

....

....

....

....

54 To what extent does human activity cause intentional rather than unintentional change to coastal landscape systems?

20 **16 marks**

Make a plan in the space below and then write your answer on a separate piece of paper.

....

....

....

....

....

Option B: Glaciated landscapes

How can glaciated landscapes be viewed as systems?

Glaciated landscapes can be viewed as systems

Glaciers are large bodies of ice able to move across, and shape, the landscape. Glaciated landscapes can be viewed as systems, comprising inputs, stores, throughputs and outputs of both energy and material. Systems can exist in a state of equilibrium, when inputs equal outputs. If this equilibrium is disturbed, the system may self-regulate and restore equilibrium; this dynamic equilibrium is an example of negative feedback.

1. Outline the three types of energy that can be inputs into glaciated landscape systems. **3 marks**

....

....

....

2. Describe one output of energy and one output of material from a glaciated landscape system. **4 marks**

....

....

....

....

....

3. What is dynamic equilibrium? **3 marks**

....

....

....

4. Distinguish between stores and flows (transfers) in a glacier system. **2 marks**

....

....

5. How is glacier mass balance calculated? **2 marks**

....

....

Glaciated landscapes are influenced by a range of physical factors

Several factors influence the way geomorphic processes work and how glaciated landscapes develop. Factors include climate, geology, latitude, altitude, relief and aspect. They can influence the operation of the glaciated landscape system by affecting the inputs, stores, throughputs and outputs. The influence of factors varies spatially and temporally. How much energy glaciers have, and how they move, greatly affects their ability to carry out geomorphic processes.

6. How do relief and aspect influence the movement of glaciers?   4 marks

....

....

....

....

7. Explain the link between altitude and climate.   3 marks

....

....

8. Explain how rock resistance is influenced by lithology and structure.   6 marks

....

....

....

....

....

....

9. Study Figure 5, which shows the intensity of glacial erosion in Britain during part of the Pleistocene glacial period. Describe and suggest reasons for the pattern of erosional intensity.   6 marks

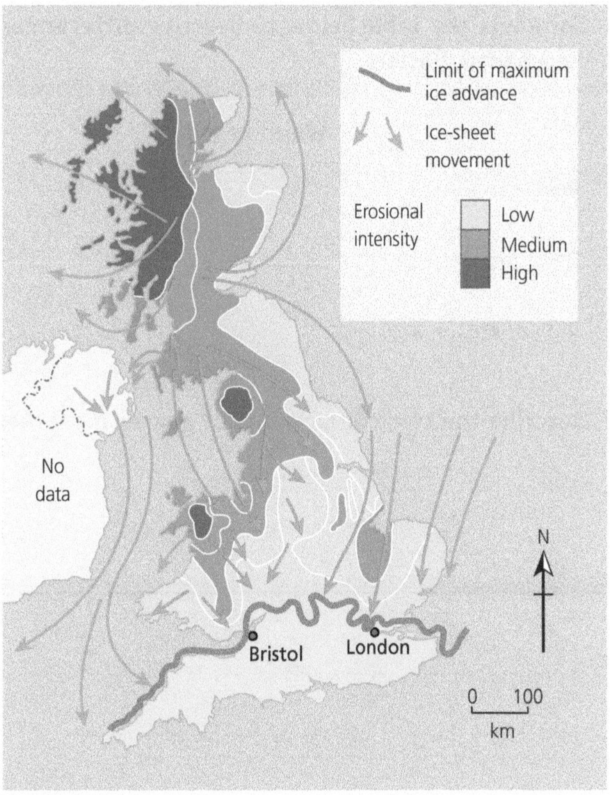

Figure 5

....

....

....

....

....

....

....

....

....

....

....

....

There are different types of glacier and glacier movement

Glacier ice forms from snow by the process of diagenesis (compression and compaction of snow due to the addition of further snowfall year on year). There are different types of glacier, including valley glaciers and ice sheets, and they have different characteristics. Warm-based and cold-based glaciers move differently; processes of movement include basal sliding and internal deformation.

10 **Describe and explain the formation of glacier ice.** 4 marks

....

....

....

....

....

....

11 **State the typical densities of:**

Fresh snow	
Firn	
Glacier ice	

3 marks

12 **What is the minimum size of an ice sheet?** 1 mark

....

13 **Complete the table below to describe differences between warm-based and cold-based glaciers.** 6 marks

	Warm-based	Cold-based
Relief		
Basal temperature		
Rate of movement		

14 **Describe the two components of glacier movement by internal deformation.** 4 marks

....

....

....

....

....

....

How are glacial landforms developed?

Glacial landforms develop through a variety of interconnected climatic and geomorphic processes

Geomorphic processes include weathering (physical/mechanical, chemical and biological), mass movement, glacial erosion, nivation, transportation and deposition. These geomorphic processes interconnect with climatic processes such as temperature fluctuations and precipitation.

15 **What does chemical weathering do to rock?** *(2 marks)*

....

....

16 **Why might freeze–thaw weathering be of limited importance in some glacial environments?** *(2 marks)*

....

....

17 **What is the key difference between slumps and linear slides?** *(2 marks)*

....

....

18 **Describe the process of glacial plucking.** *(2 marks)*

....

....

19 **State the position within a glacier where these different types of glacial transportation can occur:** *(3 marks)*

englacial ....

subglacial ....

supraglacial ....

20 **Describe and explain how sediment can be deposited by glaciers.** *(6 marks)*

....

....

....

....

....

....

Glacial landforms include:

- those that are predominantly the result of erosion, including corries, arêtes, pyramidal peaks, troughs, roches moutonnées and striations. However, other geomorphic processes may also be involved, such as weathering and mass movement (subaerial processes)
- those that are predominantly the result of deposition, including terminal, lateral and recessional moraines, erratics, drumlins and till sheets. However, other geomorphic processes may also be involved, especially transportation

21 Draw an annotated diagram to explain the formation of a corrie.   6 marks

22 Explain how striations can indicate the movement of a glacier.   2 marks

....

....

23 Explain the influence of subaerial processes on the formation of lateral moraines.   2 marks

....

....

24 Differentiate between the formation of lodgement till and ablation till.   2 marks

....

....

25 Describe the typical shape of a drumlin.   3 marks

....

....

....

26 Study Figure 6. With the aid of Figure 6, explain the formation of a roche moutonnée.   6 marks

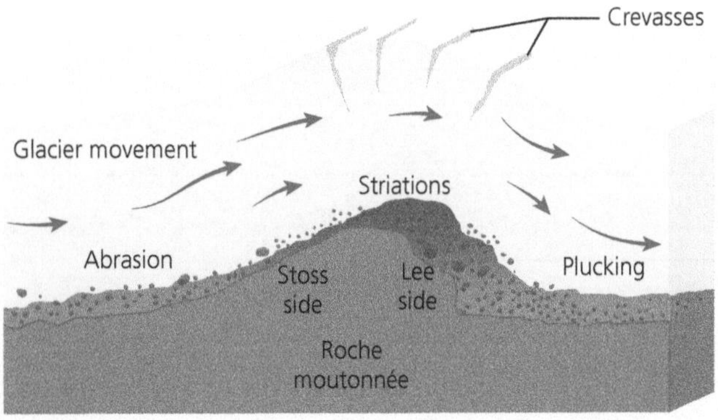

Figure 6

20

....

....

....

....

....

....

Glacial landforms are interrelated and together make up characteristic landscapes

This topic is focused on two case studies: a landscape associated with the action of valley glaciers and a landscape associated with the action of ice sheets – one of which must be from outside the UK.

For each, you should know and be able to explain:

- the physical factors influencing the formation of its landforms
- the interrelationship of the landforms within the landscape system
- how and why the landscape system changes over time, from millennia to seconds

Note: answers to these questions will depend on your chosen case studies.

27 **Describe the geology of one of your case study locations, and explain how it influences the landscape.** *6 marks*

....

....

....

....

....

....

28 **Explain how advancing ice influenced the landscape in one of your case studies.** *4 marks*

....

....

....

....

29 **Choose two depositional landforms in your case study of a landscape associated with the action of ice sheets, and explain how they are interrelated.** *4 marks*

....

....

....

....

30 Describe and explain the long-term (over millennia) changes to the landscape in one of your case studies.

6 marks

....

....

....

....

....

....

How do glacial landforms evolve over time as climate changes?

Glacio-fluvial landforms exist as a result of climate change at the end of glacial periods

Climate change occurs at the end of glacial periods, and this affects the geomorphic processes that take place. These processes produce landforms, including kames, eskers and outwash plains. These landforms can be modified by present and future climate change.

31 Study Figure 7, which shows the thickness of glacio-fluvial deposits laid down over 8 years during a glacial period in New York State. Using data from Figure 7, describe and explain the variations in the thickness of the deposits.

6 marks

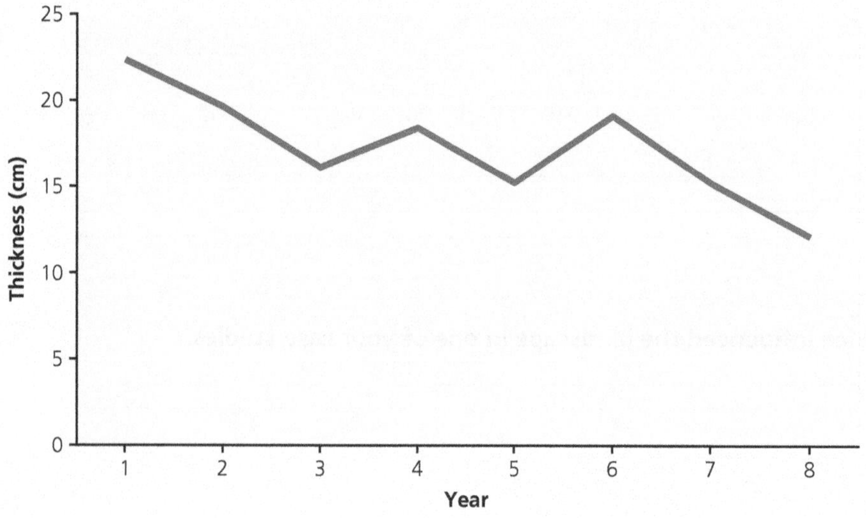

Figure 7

....

....

....

....

....

....

32 Why do glaciers often have supraglacial meltwater streams flowing along their edge? *2 marks*

....

....

33 Describe the different positions in a glacier at which delta kames originate. *3 marks*

....

....

....

34 How do supraglacial streams form kame terraces? *4 marks*

....

....

....

....

35 Describe and explain the formation of eskers. *6 marks*

....

....

....

....

....

....

....

36 Why are eskers sometimes 'beaded'? *2 marks*

....

....

37 Why are meltwater streams often braided? *3 marks*

....

....

....

38 Suggest how glacio-fluvial landforms can be modified after their initial formation. *5 marks*

....

....

....

....

....

Periglacial landforms exist as a result of climate change before and/or after glacial periods

Periglacial environments have permafrost, seasonal temperature variations and geomorphic processes dominated by freeze–thaw cycles. They are commonly located around the edges of ice sheets and glaciers.

39 Using Figure 8, describe the key climatic characteristics of Barrow, Alaska. *(4 marks)*

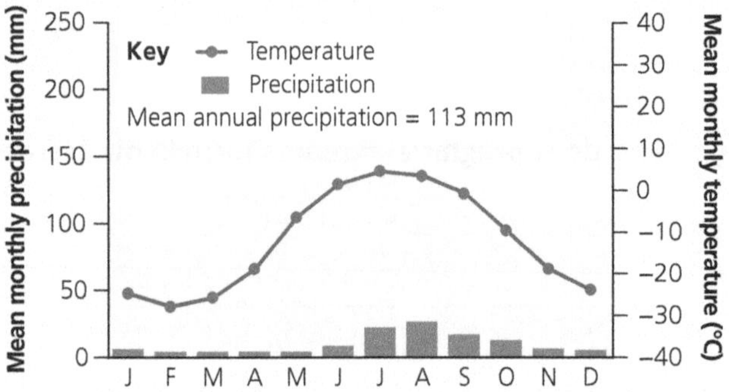

....

....

....

....

....

....

....

Figure 8

40 What is 'permafrost'? *(2 marks)*

....

....

41 Which part of the British Isles was periglacial during the maximum ice advance in the most recent glacial period? *(1 mark)*

....

42 Explain the process of frost-heave. *(3 marks)*

....

....

....

43 Distinguish between the appearance of stone stripes and stone polygons. *(2 marks)*

....

....

44 With the aid of a diagram, explain the formation of closed-system pingos. *(6 marks)*

How does human activity cause change within glacial and periglacial landscape systems?

Many glacial and periglacial landscapes have opportunities for human activity. However, such activity on any significant scale can have major impacts on these often delicately balanced landscape systems.

Human activity causes change within periglacial landscape systems

You will study one periglacial landscape that is being used by people. You should know and understand:

- the human activity taking place and the reasons for it taking place, such as resource extraction
- the impacts on processes and flows of material and/or energy through the periglacial system, such as heat produced by buildings
- the effect of these impacts in changing periglacial landforms, such as thawing of permafrost
- the consequences of these changes on the landscape, such as development of thermokarst

Note: answers to these questions will depend on your chosen case study.

45 **Draw a sketch map to show how and why your chosen periglacial landscape is being used by people.**

6 marks

46 **How has the human activity affected processes/flows of energy and materials in the system?**

5 marks

....

....

....

....

....

47 **What effects have the changes in processes/flows of energy and materials had on the landscape?**

3 marks

....

....

....

Human activity causes change within glaciated landscape systems

You will study one glaciated landscape that is being used by people. You should know and understand:

- the human activity taking place and the reasons for it taking place, such as dam construction
- the impacts on processes and flows of material and/or energy through the glacial system, such as trapping of sediment
- the effect of these impacts in changing glacial landforms, such as increased channel scour below dams
- the consequences of these changes on the landscape, such as changes to the valley floor

Note: answers to these questions will depend on your chosen case study.

48 **Explain why the human activity is taking place in your chosen glaciated landscape.** 6 marks

....

....

....

....

....

....

....

49 **Describe and explain the impacts human activity has had on processes and flows of materials and energy in your chosen glaciated landscape.** 6 marks

....

....

....

....

....

....

50 **With the aid of a diagram, show how the changes to the processes/flows have influenced the landscape.** 4 marks

Exam-style questions

51 Explain the influence of energy flows on the formation of a glacial trough. ⏱ 10 **8 marks**

....

....

....

....

....

....

....

52 Study Table 2, which shows ice accumulation in a glacier in the Andes, 2017.

Table 2

Months	Ice accumulation (cm of water equivalent)
Sept–Oct	5.5
Nov–Dec	2.4
Jan–Feb	1.8
March–April	2.1
May–June	4.4
July–Aug	6.6

a Calculate the mean value of ice accumulation shown in Table 2. You must show your working. ⏱ 3 **2 marks**

....

....

....

b Calculate the median value of ice accumulation shown in Table 2. You must show your working. ⏱ 3 **2 marks**

....

....

....

53 Study Figure 9, which shows a landscape during glaciation. Using evidence from Figure 9 to aid your answer, explain how glacio-fluvial landforms develop during postglacial periods. ⏱ 7 **5 marks**

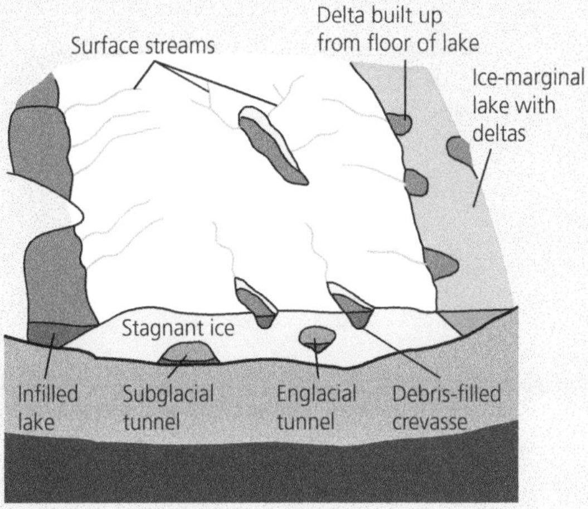

....

....

....

....

....

....

....

....

Figure 9

....

....

54 To what extent does human activity cause more significant change in periglacial landscape systems than in glaciated landscape systems?

20   16 marks

Option C: Dryland landscapes

How can dryland landscapes be viewed as systems?

Dryland landscapes can be viewed as systems

Drylands are regions where average annual evapotranspiration is significantly higher than precipitation. Dryland landscapes can be viewed as systems, comprising inputs, stores, throughputs and outputs of both energy and material. Systems can exist in a state of equilibrium, when inputs equal outputs. If this equilibrium is disturbed, the system may self-regulate and restore equilibrium; this dynamic equilibrium is an example of negative feedback.

1. Outline the three types of energy that can be inputs into dryland landscape systems. *3 marks*

2. Describe one output of energy and one output of material from a dryland landscape system. *4 marks*

3. What is dynamic equilibrium? *3 marks*

4. Distinguish between stores and flows (transfers) in a dryland system. *2 marks*

5. How is aridity index calculated? *2 marks*

Dryland landscapes are influenced by a range of physical factors

Several factors influence the way geomorphic processes work and how dryland landscapes develop. Factors include climate, geology, latitude, altitude, relief, aspect and availability of sediment. They can influence the operation of the dryland landscape system by affecting the inputs, stores, throughputs and outputs. The influence of factors varies spatially and temporally.

The amount of energy that water and wind have greatly affects their ability to carry out geomorphic processes.

6 **How do relief and aspect influence the location of dryland landscapes?** 4 marks

7 **Explain the link between altitude and aridity.** 3 marks

8 **Explain how rock resistance is influenced by lithology and structure.** 6 marks

9 **Suggest why sediment availability varies from place to place.** 4 marks

There are different types of dryland

There are three main types of dryland: polar drylands, mid- and low-latitude deserts and semi-arid environments.

10 Study Figure 10, which shows the global distribution of mid- and low-latitude deserts.

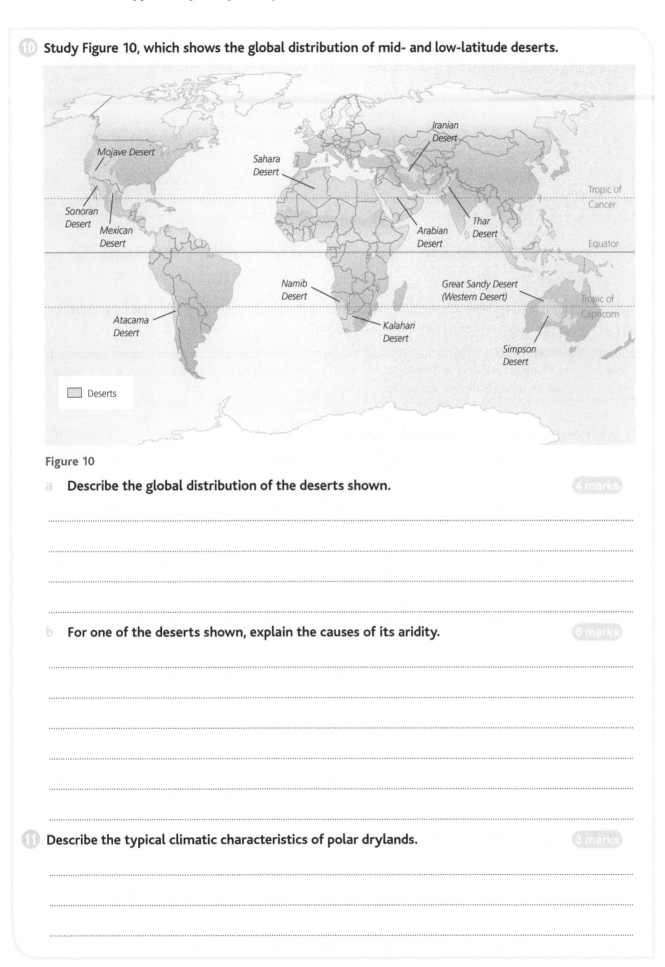

Figure 10

a   Describe the global distribution of the deserts shown.　4 marks

b   For one of the deserts shown, explain the causes of its aridity.　6 marks

11 Describe the typical climatic characteristics of polar drylands.　3 marks

31

12. What is the aridity index for semi-arid environments? *2 marks*

....

....

13. Copy and complete the table below to describe the landscapes of the different drylands. *9 marks*

	Polar drylands	Mid- and low-latitude deserts	Semi-arid environments
Ground conditions			
Vegetation			
Drainage			

How are landforms of mid- and low-latitude deserts developed?

Dryland landscapes develop because of a variety of interconnected climatic and geomorphic processes

Geomorphic processes include weathering (physical/mechanical, chemical and biological), mass movement, fluvial and aeolian erosion, transportation and deposition. These geomorphic processes interconnect with climatic processes such as temperature and precipitation.

14. What does mechanical weathering do to rock? *2 marks*

....

....

15. Why might biological weathering be of limited importance in some drylands? *2 marks*

....

....

16. Distinguish between rock falls and debris flows. *4 marks*

....

....

....

....

17. What is meant by the term 'ephemeral river'? *2 marks*

....

....

18. State three reasons why dryland rivers may have high levels of energy. *3 marks*

....

....

....

19 Describe how wind transports sediment. *6 marks*

....

....

....

....

....

....

20 Explain how aeolian deposition may involve positive feedback. *3 marks*

....

....

....

Dryland landforms include:

- those that are predominantly the result of erosion, including wadis, canyons, pedestal rocks, ventifacts and desert pavements. However, other geomorphic processes may also be involved, such as weathering and mass movement (subaerial processes)
- those that are predominantly the result of deposition, including barchans, linear dunes, star dunes, alluvial fans and bajadas. However, other geomorphic processes may also be involved, especially transportation

21 Draw an annotated diagram to explain the formation of a pedestal rock. *6 marks*

22 Explain how ventifacts can indicate the dominant wind direction. *2 marks*

....

....

23 Explain the influence of subaerial processes on the formation of canyons. *4 marks*

....

....

....

....

24 Distinguish between an alluvial fan and a bajada. 3 marks

....

....

....

25 How is the alignment of linear dunes related to the prevailing wind direction? 1 mark

....

26 Annotate Figure 11 to explain the formation of a barchan. 6 marks

(a) In plan   (b) In profile

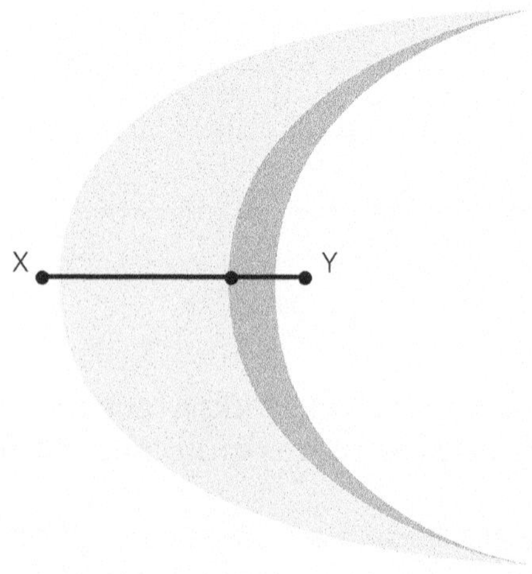

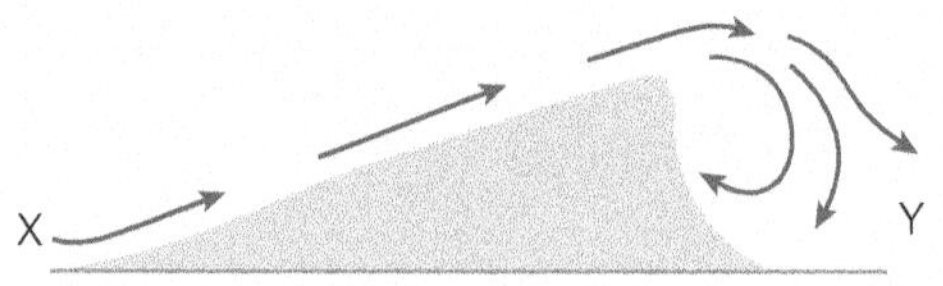

Figure 11

Dryland landforms are interrelated and together make up characteristic landscapes

This topic is focused on two case studies: a mid-latitude desert and a low-latitude desert. For each, you should know and be able to explain:

- the physical factors influencing the formation of its landforms
- the interrelationship of the landforms within the landscape system
- how and why the landscape system changes over time, from millennia to seconds

Note: answers to these questions will depend on your chosen case studies.

27 Draw a sketch map or diagram of your low-latitude desert case study location to show the main features of the landscape. 6 marks

28 Choose two landforms in your case study of a low-latitude desert and explain how they are interrelated. (4 marks)

....

....

....

....

29 Describe the geology of one of your case studies, and explain how it influences the landscape. (6 marks)

....

....

....

....

....

....

30 Explain how sediment availability influences the landscape in one of your case studies. (4 marks)

....

....

....

....

31 Choose two landforms in your case study of a mid-latitude desert and explain how they are interrelated. (4 marks)

....

....

....

....

32 Describe and explain the short-term changes to the landscape in one of your case studies. (6 marks)

....

....

....

....

....

....

How do dryland landforms evolve over time as climate changes?

Fluvial landforms can exist in dryland landscapes as a result of earlier pluvial periods

The climate in drylands has oscillated between wet and dry conditions over the last 500,000 years. Many landforms, such as inselbergs and pediments, have been influenced by fluvial processes that occurred in wetter, pluvial periods.

33 **For a specific pluvial period in a dryland, describe its climatic conditions.**　3 marks

....

....

....

34 **Explain one reason why a dryland may have experienced an earlier pluvial period.**　3 marks

....

....

....

35 **Give one piece of evidence that indicates that a dryland experienced an earlier pluvial period.**　2 marks

....

....

....

36 **With the aid of an annotated diagram, explain the formation of an inselberg.**　6 marks

37 Study Figure 12. Suggest how fluvial processes have influenced the formation of the pediment.

(4 marks)

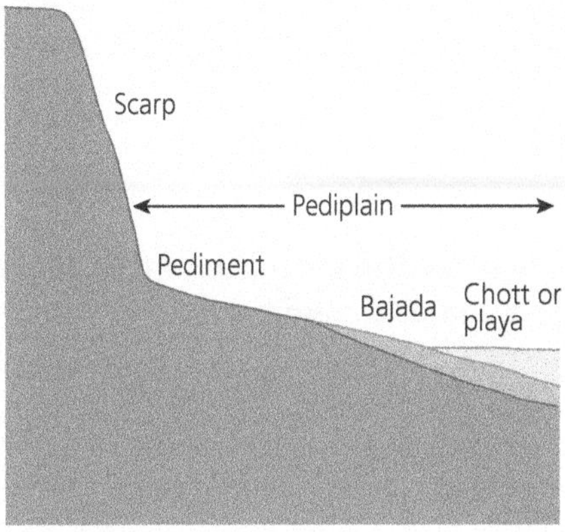

....

....

....

....

....

....

....

Figure 12

38 Suggest how inselbergs may be modified by future climate change.

(4 marks)

....

....

....

....

Periglacial landforms can exist as a result of earlier colder periods

The global climate has fluctuated between colder and warmer periods on numerous occasions in the past 500,000 years. Periglacial environments have permafrost, seasonal temperature variations and geomorphic processes dominated by freeze–thaw cycles. Many landforms such as frost-shattered debris, nivation hollows and solifluction deposits were formed during periglacial conditions in these colder periods.

39 Study the photo below. Describe the appearance of the debris in the foreground.

(4 marks)

40 Suggest how this debris was formed. 4 marks

....

....

....

....

41 Explain what the presence of this debris in a dryland landscape indicates about its past climatic conditions. 2 marks

....

....

....

....

42 How might these rocks be modified by present and future climate changes? 4 marks

....

....

....

....

43 Explain why solifluction deposits are found in present-day dryland landscapes. 6 marks

....

....

....

....

....

....

44 With the aid of a diagram, explain the formation of a nivation hollow. 6 marks

How does human activity cause change within dryland landscape systems?

Water supply issues can cause change within dryland landscape systems

You will study one dryland landscape that is being used by people. You should know and understand:

- the water supply issue taking place and the reasons for it, such as water shortage due to drought
- the impacts on processes and flows of material and/or energy through the dryland landscape system, such as trapping of sediment behind dams or modifying river flows
- the effect of these impacts in changing dryland landforms, such as decreased growth of wadis
- the consequences of these changes on the landscape, such as reducing depositional landforms or slowing pediment development

Note: answers to these questions will depend on your chosen case study.

45 **Draw a sketch map to explain why your chosen dryland landscape is being used by people for water supply.**
6 marks

46 **How has the human activity affected processes and flows of energy and materials in the system?**
3 marks

....

....

....

47 **What effects have the changes in processes/flows of energy and materials had on the landscape?**
3 marks

....

....

....

Economic activity causes change within dryland landscape systems

You will study one dryland landscape that is being used by people. You should know and understand:

- the economic activity taking place and the reasons for it, such as tourism
- the impacts on processes and flows of material and/or energy through the dryland landscape system, such as damage by dune buggy use
- the effect of these impacts in changing dryland landforms, such as increased erosion rates on dunes
- the consequences of these changes on the landscape, such as increased loess accumulation

Note: answers to these questions will depend on your chosen case study.

48 **Explain why the economic activity is taking place in your chosen dryland landscape.**   6 marks

....

....

....

....

....

....

49 **Describe and explain the impacts human activity has had on processes and flows of energy and materials in your chosen dryland landscape system.**   6 marks

....

....

....

....

....

....

50 **With the aid of a diagram, show how changes to the processes and flows have influenced the landscape.**   4 marks

Exam-style questions

51 Explain the influence of energy flows on the formation of a desert pavement. ⏱ 10 **8 marks**

....

....

....

....

....

....

....

52 Study Table 3, which shows sediment accumulation on a sand dune in the Sahara in 2017.

Table 3

Months	Sediment accumulation (m³)
Sept–Oct	5.5
Nov–Dec	2.4
Jan–Feb	1.8
March–April	2.1
May–June	4.4
July–Aug	6.6

a Calculate the mean value of sediment accumulation shown in Table 3. You must show your working. ⏱ 3 **2 marks**

....

....

....

b Calculate the median value of sediment accumulation shown in Table 3. You must show your working. ⏱ 3 **2 marks**

....

....

....

....

53 Study Figure 13, which shows a dryland during a pluvial period.

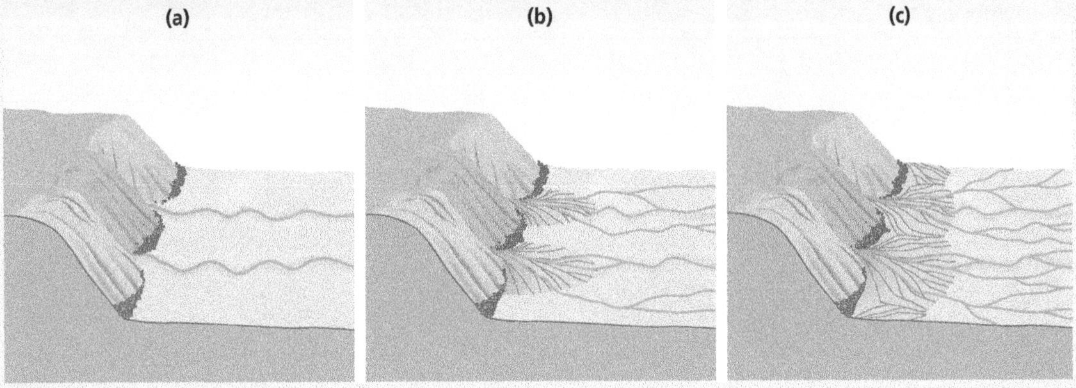

(a)   (b)   (c)

Figure 13

Using evidence from Figure 13 to aid your answer, explain how and why bajadas develop during pluvial periods. ⏱ 7 **5 marks**

....

....

....

....

....

54 To what extent does human activity related to water supply issues cause more significant change in drylands than economic activity?

20   16 marks

Changing spaces; making places

What's in a place?

Places have several characteristics and are shaped by changing flows and connections

Places are locations that are considered as having something distinctive about them. Every place has a particular place identity made up of several characteristics, including physical geography, built environment, demographic, socioeconomic, cultural and political. These characteristics are shaped by dynamic flows such as people, money and connections with other places that change over time and in their scale. You will undertake research into two contrasting local places so that you are able to write focused, detailed descriptions and analyses of their place identities.

1. State two physical characteristics of a place's identity. **2 marks**

....

....

2. State two demographic characteristics of a place's identity. **2 marks**

....

....

3. What is meant by the built environment of a place? **2 marks**

....

....

4. Contrast your two local places in terms of their present-day socioeconomic characteristics. **6 marks**

Name of place 1: _____	Name of place 2: _____

5. Explain the role that past connections have played in forming the place identity of one of your chosen local places. **4 marks**

....

....

....

....

6　Suggest how shifting flows of money and investment have influenced the place profile of one of your chosen local places.

4 marks

....

....

....

....

7　Assess the impacts that local and national politics can have on a place's identity.

6 marks

....

....

....

....

....

....

....

How do we understand place?

People see, experience and understand place in different ways and this can also change over time

People vary in the ways they view the same place. These views are subject to a person's perceptions, which are influenced by factors such as age, gender, sexuality, ethnicity, religion or role such as mother, husband, sister or employee. Perceptions can alter over time as experiences, knowledge and understanding change.

8　How might changes in life-cycle stage* influence perceptions of residential preferences?

4 marks

*Life-cycle stage describes the progress of a person through various stages based on age and family unit (e.g. young adult, family with children, friends sharing accommodation, retired)

....

....

....

....

9　Examine how male and female attitudes towards places have changed through time.

4 marks

....

....

....

....

10 Identify two places that have religious meanings for some people. **2 marks**

....

....

11 Explain the significance of emotional attachment to place for a group such as the Kurds or Basques. **6 marks**

....

....

....

....

....

....

12 What is meant by 'globalisation'? **2 marks**

....

....

13 With reference to Figure 14, describe and explain how networks and flows in the past can influence place identity today. (Answer lines on p. 46.) **6 marks**

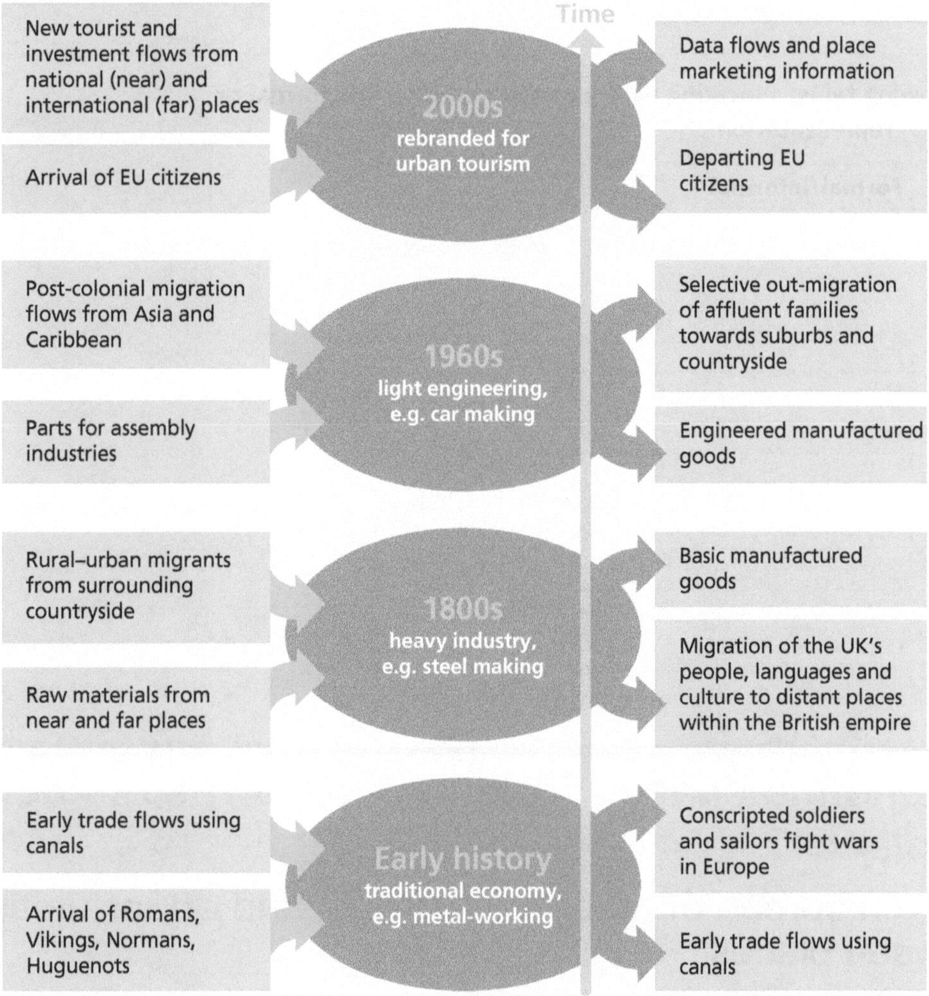

Figure 14

....

....

....

....

....

....

Places are represented through a variety of contrasting formal and informal ways

Places can be described in various ways that can be placed in one of two categories: formal or informal. Different types of representation give contrasting perspectives of a place.

14 **Distinguish between 'formal' and 'informal' representation of a place.**   *4 marks*

....

....

....

....

15 **Complete the following table, allocating each example to either the formal or informal category of representation.**   *6 marks*

Representation	Formal/informal
Government census	
Police crime data	
Novel	
Photograph	
Health data	
Blog	

How does economic change influence patterns of social inequality in places?

There is an uneven spread of resources, wealth and opportunities within and between places

Factors such as income, educational opportunities and housing interact to create social inequalities which can be measured in a variety of ways. Patterns of social inequality result in spatial inequalities, resulting in places offering contrasting standards of living and quality of life.

16 State and explain the difference between 'poverty' and 'deprivation'.   4 marks

....

....

....

....

17 Name one example of each of the following methods to investigate patterns of social inequality: 2 marks

a   cartographic

....

b   statistical

....

18 Study Table 4, which shows average life expectancy at birth (male + female) for selected UK urban places in 2018.

Table 4

Urban place	Average life expectancy (years)
Kensington + Chelsea	83.7
St Albans	82.0
Oxford	80.6
Sunderland	77.2
Blackburn	76.2
Glasgow	73.4

Using Table 4, suggest how each of the factors below might influence the contrasts in life expectancies.   8 marks

Factor	Influence on life expectancy
Income	
Employment	
Housing	
Health	

47

Economic change creates both advantages and disadvantages for people

Globalisation has led to a global shift in the location of manufacturing industries in particular. Economic structures of countries, regions and urban locations have changed, with a variety of impacts on people and places. Governments have been involved in these changes because of the effect on inequalities of economic change.

19 **Study Figure 15, which shows changes in employment structure in England and Wales, 1921–2018.**

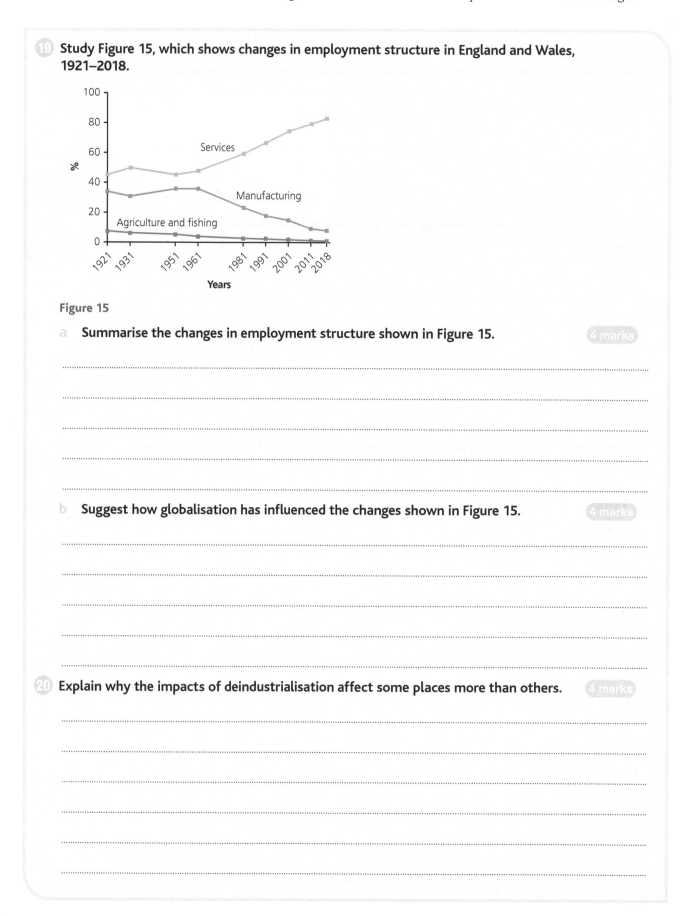

Figure 15

a **Summarise the changes in employment structure shown in Figure 15.**   4 marks

....

....

....

....

....

b **Suggest how globalisation has influenced the changes shown in Figure 15.**   4 marks

....

....

....

....

....

20 **Explain why the impacts of deindustrialisation affect some places more than others.**   4 marks

....

....

....

....

....

....

21 Outline ways in which cycles of economic change (Kondratieff cycles) bring both opportunities and inequalities for people. [6 marks]

....

....

....

....

....

....

22 State and describe two methods governments may employ to tackle inequalities. [4 marks]

Method 1

....

....

Method 2

....

....

The contrasts in social inequality between two places

This topic is focused on case studies of two places that illustrate how social inequality impacts people and places in different ways. For Questions 23 to 25 make notes in the tables below and then write full prose answers on separate sheets of paper.

23 Describe the locations of your two contrasting places. [4 marks]

Place 1 name: _____	Place 2 name: _____

24 Outline the socioeconomic conditions existing in each place using the headings below. [6 marks]

Income and employment   Health   Education

Place 1 name: _____	Place 2 name: _____

25 Outline the environmental conditions existing in each place. *4 marks*

Place 1 name: _____	Place 2 name: _____

Who are the players that influence economic change in places?

Places are influenced by a range of players operating at different scales

Public or private individuals, groups or formal organisations that influence or are influenced by the processes of change in a place are known as players. Their interactions are important in making a place's identity. You will study one country or region that has been affected by changes in its economic structure.

26 Name one example of each of the following types of player: *3 marks*

Public player regional or national scale ....

Private player global scale ....

Private player local scale ....

27 Describe the changes in economic structure experienced by your chosen place. *6 marks*

....

....

....

....

....

....

28 Identify and explain the role of two different players in generating change in the economic structure of your chosen place. *6 marks*

....

....

....

....

....

....

29 Outline the socioeconomic impacts of economic change on people in your chosen place. *3 marks*

....

....

....

How are places created through place-making processes?

Place is produced in a variety of ways at different scales

Place-making refers to the strategies used by a range of players to re-shape the environment and economy of a place to give it a renewed identity. This process usually follows a period of decline in a place, often the consequence of loss of employment such as in agriculture or industries such as iron and steel, textiles or mining.

30 Describe how architects and planners can re-design the physical characteristics of a place as part of the place-making process. *4 marks*

....

....

....

....

....

31 What is meant by the '24-hour city'? *2 marks*

....

....

32 Outline how community groups can shape a local place. *4 marks*

....

....

....

....

....

Rebranding aims to re-image and regenerate a location to give it a new place identity

A range of strategies is used to rebrand a place so that the location develops characteristics that are more likely to offer higher standards of living and quality of life to residents. Sometimes this process involves reworking and modernising the existing place identity, sometimes a new identity is created.

33 What is meant by the 're-imaging' of a place?

2 marks

....

....

....

34 Referring to named examples, explain how sport, art, heritage, retail, architecture and food can help to change place identity.

12 marks

Factor	Influence on place identity
Sport	
Art	
Heritage	
Retail	
Architecture	
Food	

35 Why can rebranding cause conflict among players involved in the process? 6 marks

....

....

....

....

....

....

....

....

Making a successful place requires planning and design

You will study a place that has undergone rebranding. You need to understand why it required rebranding, the strategies used to rebrand, how perceptions of the place were altered and assess how successful the rebranding has been.

For Questions 36 to 38 make notes in the tables below and then write full prose answers on separate sheets of paper.

36 Outline why your chosen place required rebranding. 4 marks

Name of place: _____	Reasons for rebranding
	....
	....
	....
	....
	....

37 Identify and explain the role of the strategies used in the rebranding process. 6 marks

Strategy	Role in rebranding
	....
	....
	....
	....
	....
	....

38 Consider the relative success of the rebranding as regards (a) local people, (b) players such as governments and (c) people from outside the place.

6 marks

Group	View about the relative success of the rebranding project
Local people	
Players, e.g. governments, TNCs	
People from outside the place	

Exam-style questions

39 Study the photograph below, of part of central New York, USA. Use one piece of evidence from the photo to explain how place identity is shaped.

4   **3 marks**

40 Explain the role of two players in driving economic change in a place you have studied.

(8) **6 marks**

....

....

....

....

....

....

....

....

....

41 Study Figure 16, which shows migration flows of 30–39-year-olds between selected UK cities, June 2014 to June 2015.

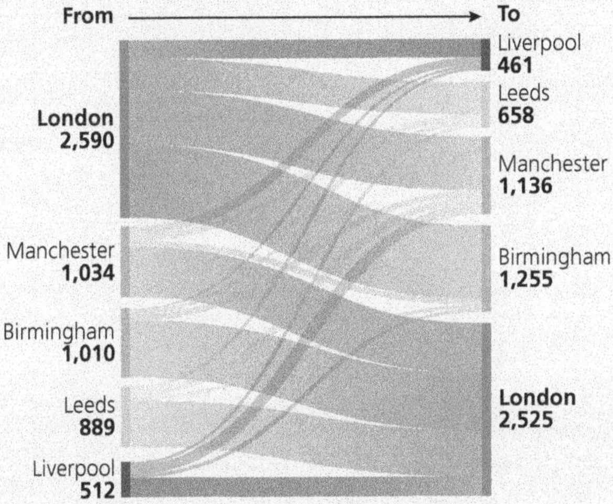

From ——————————→ To

London 2,590

Manchester 1,034

Birmingham 1,010

Leeds 889

Liverpool 512

Liverpool 461

Leeds 658

Manchester 1,136

Birmingham 1,255

London 2,525

Sources: DCLG, ONS, HM Land Registry, Colliers

Figure 16

Using evidence from Figure 16, suggest how the migration of 30–39-year-olds might affect different urban places.

(10) **8 marks**

....

....

....

....

....

....

....

....

....

....

42 To what extent does the successful making of a place require removal of past characteristics?

20 | 16 marks

Orders: please contact Hachette UK Distribution, Hely Hutchinson Centre, Milton Road, Didcot, Oxfordshire, OX11 7HH. Telephone: (44) 01235 827827. E-mail education@hachette.co.uk
Lines are open from 9 a.m. to 5 p.m., Monday to Friday. You can also order through our website: www.hoddereducation.co.uk

ISBN: 978-1-5104-5841-3

© Peter Stiff, Helen Harris and Andy Palmer 2019

First published in 2019 by
Hodder Education,
An Hachette UK company
Blenheim Court
George Street
Banbury
Oxfordshire OX16 5BH

www.hoddereducation.co.uk

Impression number   10 9 8 7 6 5 4 3 2

Year   2023 2022 2021

Cover photo: © mozZz–stock.adobe.com

Photo credits: p.37 Paul Vinten/Adobe Stock; p.54 Tierney/Adobe Stock

Typeset by Aptara, India

Printed in UK

A catalogue record for this title is available from the British Library.

HODDER EDUCATION

t: 01235 827827
e: education@bookpoint.co.uk
w: hoddereducation.co.uk

ISBN 978-1-5104-5841-3

9 781510 458413